IMAGES OF ASIA

A Garden of Eden

Other titles in the series

At the Chinese Table

Chinese Jade

Indonesian Batik: Processes, Patterns and Places

The Kris: Mystic Weapon of the Malay World

Macau

Sailing Craft of Indonesia

A Garden of Eden

Plant Life
in South-East Asia

W. VEEVERS-CARTER

SINGAPORE
OXFORD UNIVERSITY PRESS
OXFORD NEW YORK
1986

Oxford University Press

Oxford New York Toronto
Petaling Jaya Singapore Hong Kong Tokyo
Delhi Bombay Calcutta Madras Karachi
Nairobi Dar es Salaam Cape Town
Melbourne Auckland

and associates in
Beirut Berlin Ibadan Nicosia

OXFORD is a trademark of Oxford University Press

ISBN 0 19 582658 2

Printed in Singapore by Koon Wah Printing Pte. Ltd.
Published by Oxford University Press Pte. Ltd.,
Unit 221, Ubi Avenue 4, Singapore 1440

Contents

I

Introduction

SOUTH-EAST ASIA is an entity only in a broad geographical sense. Politically, Thailand, Laos, Cambodia/Kampuchea, Vietnam, Malaysia, the Philippines and Indonesia are usually considered South-East Asian countries, with Burma an aloof but potential candidate to the west, and Papua New Guinea, the Solomons and Hong Kong less frequently included to the east and north. Physically, there are clearly great differences between continental mainland South-East Asia, where volcanic formations are relatively rare, and the islands which owe their shape or even existence to volcanic action, and which form such an easily traced arc of the Pacific's 'ring of fire'. Though generally hot, where a continental type of climate prevails there are marked seasonal differences in rainfall and some in temperature. The western countries are subject to Indian Ocean monsoon winds and rain; the south-easternmost to hot dry winds blowing north from the Australian deserts. In between, however, the insular and mountainous nature of Malaysia, Indonesia and the Philippines, together with their position astride the Equator, ensures that these countries are both uniformly hot and uniformly humid, a climatic type that has apparently remained unaltered since the beginning of the Cretaceous period, about 120 million years ago. Only in such stable conditions could South-East Asia's immensely tall rain forests, her botanic crowning glory, have developed. Apart from volcanic eruptions, and Borneo has mostly escaped even these, all large-scale disturbances—extensive commercial logging, clearance for transmigration schemes, the unbelievable fires which raged in East Kalimantan in 1983—have been the result, directly or indirectly, of recent human interference; the long-term results of such damage can only be guessed.

A generally warm, humid climate with little variation in temperature favours a luxuriant variety of vegetation even outside the hothouse confines of the rain forest. Warmth and water for all, hardship for none, might be the landlord's advertisement; the plants take advantage of it—and South-East Asians take full advantage of their plants, and their knowledge of their qualities. In spite of some westernization, South-East Asian nations retain their dependence on their agricultural resources, including those of their dwindling forests, to an amazing degree. The nutrition of all depends on the cultivation of rice; bamboo in some form is used by all; 80 per cent of the fruits consumed are native and most are rain forest in origin; orchid culture, with constant recourse to wild species in the forests, is a huge industry, as is the making of furniture from rattan, another rain forest product. Hardwoods, oils, resins, fibres, spices, tannins, dyes and a whole pharmacopoeia of medicinal plants: the list is long. In no other part of the world is there so dense a population making such intensive use of such a variety of forest products irrespective of their level of cultural 'modernization'. This close dependence on the environment will hopefully encourage South-East Asia to avoid the more destructive and barren forms industrial development has taken in the West.

In this vegetative paradise, the acute observer will have noticed that not all the common species are local in origin in spite of the vaunted plenty of the rain forests. The coastal vegetation, for instance, bears a remarkable resemblance to the vegetation on similar shores throughout the Indo-Pacific area. Plants with edible fruits or beautiful flowers have obviously travelled widely within the tropic zones, the ability to survive in human environments being an important qualification for such an artificial distribution. If South-East Asia is the original source of many plants now known and used throughout the world—the banana is an outstanding example—then it has also become home to many a foreign species. South-East Asians may be more aware than most peoples of the contribution plants make to the earthly quality of life.

2

The Sandy Seashore

FROM the shores of East Africa to the islands of the western Pacific, the traveller is likely to recognize some 'old friends' growing along the sandy beaches, just as the sight of mangroves will be familiar in the muddy estuaries. There are those velvety silver-grey leaves again and the sweet-smelling flowers of *Tournefortia*, here on a low tree the hollyhock-like yellow blooms of the *Hibiscus tiliaceous*, one of the most ubiquitous and useful species of the hibiscus family. The coconut palm is everywhere, and nearly as often the pandanus or screw pine, easily recognizable by their long bladed leaves, stilt roots and handsome orange fruits. Spiky spinifex and bright green goat's foot creeper with its morning glory type of flower grow in the sand just above high water; the contorted iron-hard roots of the *Pemphis acidula* shrub with its small leaves and small white flowers favours a broken scree of sandstone or coral. The common straggly plant generally sheltering a hum of mosquitoes and producing not very many long shiny leaves on its pithy stems looks good for nothing, but forms a surprisingly good windbreak: this is *Scaevola*. And then there are the big trees. Casuarinas, tall and shaggy, stand majestically at the edge of the dunes, seemingly impervious to the salt winds. The tree with some bright red leaves and horizontal branches will be the Indian country almond; the producer of the big square fibrous-husked fruit is *Barringtonia*, and the giant with the big recumbent limbs is *Calophyllum*.

Casuarina (ru or aru)

The Casuarina (fig. 1), an Australian tree, is not a pine, though because of its habits and appearance many people mistake it for

4

Figure 1. Young casuarina, showing habit and 'needle'-like leaves

one. It is single-stemmed like a pine and an equally good pioneer on bare sandy soils; the bark is shaggy and rough, splitting off easily like that of some pines; they even seem to have cones and needles, and the litter these make when they have dried and dropped off the tree is remarkably similar to the beds of pine needles in a pine forest. Pine needles, however, are the actual leaves of the pine tree compressed to a fine cylinder, whereas the 'needles' of casuarinas are slender twigs on whose segments tiny scale-like leaves are just visible if you look closely. In both cases, these modified structures reduce water loss to a minimum and are useful adaptations for trees which can colonize such poor soils and exposed sites.

As for the 'cones', these are the fruit or seed heads, which look like minute round green durians before they ripen. On ripening, each protuberance splits open to release the small light wind-borne seed; the empty fruit then drops to join the piles of dead twigs at the base of the tree.

Where the mountain casuarina grows, the amount of dry matter around the trees constitutes a serious fire hazard, but though other trees may suffer, the casuarina after a certain age is extremely resistant to fire and will even sprout strongly again from a blackened stump.

Country Almond (ketapang)

The Indian country almond, *Terminalia catappa* or *ketapang* (Plate I), is one of the most common, and most commonly planted trees throughout the Old World tropics. Not confined to the coast, but often found there, it is an easy tree to recognize because of its distinctive style of branching and for the fact that, almost alone among tropical trees, the dying leaves turn to a brilliant scarlet. The special form of branching is called, after this species, 'Terminalia branching' or, descriptively, 'pagoda branching'. The lead shoot grows upward strongly then several nearly horizontal side branches form one of which then takes over as the lead shoot. Since the leaves all grow in tufts on the upper sides of the branches, the effect of intermittent growth and spread is indeed pagoda-like, though in the Chinese or Japanese idiom rather than the Burmese. Once the side branches have themselves branched out and become covered with leaves, like the old tree pictured in Plate I, the effect is not so noticeable, but the flat lower layer remains and makes them excellent shade trees.

The fruit, pinkish when it first drops from the tree and the fibrous husk still contains a sweetish juice, is a great favourite with children and pigs, the reward in the centre being the two almond-flavoured slivers. But few except children have the patience to de-husk these fruit just for the nuts; a small bowlful is at least an hour's steady labour.

Barringtonia (putat laut)

The *Barringtonia asiatica* or *putat laut* (Plate II) is so common to Indo-Pacific shores that the whole 'formation' of such plants as commonly grow on these shores is named after it: the 'Barringtonia formation'. Although the Barringtonia's square-based pointed seeds, called by the Seychellois 'bonnet carré', are quite unmistakable and the beautiful but ephemeral flower easy to recognize, the leafy tree can be confused with both the *ketapang* or country almond and the *buah*, or *Fagraea crenulata*. Barringtonia leaves wither yellow or pale orange, however, never red, and when young are a pale olive. They grow in rosettes and are big, fleshy and distinctly pointed; *Fagraea* leaves are even larger, rather cabbage-like and grow in alternating pairs.

The thick husk of the Barringtonia seed, like that of the dry coconut, can float in salt water for a long time before becoming water-logged—one reason for their wide maritime distribution.

Calophyllum (bitangor)

The other large tree of the seashore is the *Calophyllum inophyllum* or *bitangor*. Sometimes also called the *penaga laut* or 'hardwood (of the) sea' because of the excellent quality of its timber; it is a favourite ship-building wood wherever it grows, not the least of its advantages being the huge branches, often 10 m or more long, which lie along the sand so conveniently just above high water. *Kalos* means beautiful in Greek and the leaves are indeed very beautiful, a shiny brilliant green with exceptionally fine and close nearly horizontal veining. The seeds are round, about 3 cm in diameter and slightly pointed at one end; the flowers white and sweetly scented. The rugged and deeply fissured bark is as distinctive as the leaves—in all, a hard tree to miss once one has seen a few.

Pandanus (pandan)

Pandanus or screw pine, though often found along the beaches, is also a plant of lowland river banks and can take many forms, from species with short trunks and long leaves to the tall tree–like forms with their leaves growing in tufts at the ends of branches like the one in Plate III. Both trunks and leaves come in armed and unarmed versions but all species have the characteristic stilt roots growing out from the base of the tree above ground level and some have aerial roots as well which drop down from the lower branches, possibly in response to the wide variation in water level of some lowland streams. The fruits are also characteristic of the genus: a round or oval composite fruit made up of distinct but attached segments, usually orange or red when ripe and roughly similar in size and looks to a pineapple.

The leaves of many species are viciously armed, with two or sometimes four rows of razor-sharp recurved hooks. When exploring coastal thickets or streams where the leaves hang down over the water and bar the way, the unwary can end up with bad lacerations even by only accidentally brushing against a leaf. But the usefulness of these long strap-like leaves far outweighs their disadvantages, and throughout the range of this genus it is for their leaves that these plants are so highly valued. A skilfully wielded knife disposes of the hooks; the leaves are then retted, like sisal, and cut into coarse or fine wide or narrow strips—according to the use planned—whose virtue is that they remain soft and flexible even after drying, unlike coconut and bamboo, while being intrinsically more durable than grass fibres or raphia. From Mauritius to Polynesia, a wide variety of hats, table mats, shopping bags, sleeping mats and sugar or coffee sacks are made from various pandanus leaves, and they always command the best prices in the local markets.

3
The Mangrove Swamp

A visitor to South-East Asia might not choose to spend much, if any, time in a mangrove swamp. In the first place, mangroves, although there are many species, resemble each other more than they appear different, since their habitat governs their general appearance. Nevertheless, they have several remarkable qualities, not the least of which is the shelter they provide for a wide variety of birds, amphibian and aquatic fauna. They are also a highly specialized plant form and therefore highly efficient. Their seeds drop into the water already sprouted, ready to root between tides. They have special breathing roots which emerge above the level of the mud to maintain their supplies of oxygen. Their supporting stilt roots trap the silt carried down by the rivers, feeding from this rich supply of nutrients while at the same time preventing the soil from being washed out to sea. At the same time the seaward roots break up the action of the waves while preventing the larger marine predators from penetrating the shelter they create for the fry of several fish species as well as for prawns and crabs. The leaves and bark shed by mangroves add to the nutritious soup on which such warm water marine species feed. Their upper branches provide food and nest sites for herons, storks and ibis and, in Borneo, for the large red and silver proboscis monkeys, the species whose males have such long and pendulous noses. Taking a boat among the mangrove swamps or through the lowland riverine swamp forests is probably the only way the ordinary traveller can hope to catch sight of these magnificent creatures.

In a word, although mangroves may look uninteresting, they are the great creators of new land, moving ever seaward as the mud builds up and solidifies behind them because they can only live within reach of the tides, and their role as a nursery area for valuable marine species cannot be underestimated. Because they catch downriver silt, however, mangroves are peculiarly sensitive to pollution, so that even when they are not being directly exploited as a source of hardwood poles, tanbark, firewood or charcoal for the coastal cities—which have been their chief uses for thousands of years—they could still be destroyed by modern factories releasing effluents upstream. Such pollution would also, of course, affect the many crustacea and fish species spawning and living among them.

An example of what can happen is provided by the case of the important Paddyland Extension Scheme in Lower Burma, financed by the World Bank. This project has already cut and plans to cut more of the still large mangrove swamps of the Irrawaddy Delta for conversion to rice fields. This may be a commercial as well as an ecological mistake. Apart from the fact that nearly all the firewood and charcoal for the cities of Rangoon, Bassein, Mergui and Mandalay come from the delta mangroves, employing large numbers of people, another large section of the local population makes its living in the fish and prawn industries. Prawns of various sizes are a basic ingredient of Burmese and Thai cuisines, and prawns are one of Burma's most valuable exports. In cutting mangroves to farm rice, important sources of employment, protein and foreign exchange are being diminished in favour of increasing low value starch production.

4
Fruits of the Rain Forest

SOME botanists think the swampy flood zones of tropical rivers contain the clues to how the flowering forest was pioneered. The importance of water to all plants confirms their aquatic origin, and such swamps are still the home of spiny thick-trunked palms, *pandanus*, banana relatives and the thick stem-med water lilies, the massively constructed sort of plant predicated as the primitive broad-leafed tree's ancestral type. There is, however, as E. J. H. Corner says in *The Life of Plants* a missing link in the plant world as well as one between human and simian ancestors. Conifers evolved before the flowering plants but were by no means the first lofty trees. The earlier 40 m high spore-producing relatives of the club-mosses must once have resembled the Araucaria forests of New Guinea. But from this pre-Cretaceous era, only the cycads and tree ferns remain, their thick stocky stems with rosettes of leaves at the top the archetypal shape of the primitive tree.

Judging by these plants there was little food in such forests for animal life, nor do coniferous forests offer much in comparison to those composed of broad-leafed trees. Flowering plants produce nectar and fruits for animals as a means of survival, using animals as pollinators and seed dispersers, and as the processors of seed casings, dead leaves, shed bark and branches into reabsorbable nutrients. Both animate and inanimate partners benefit from the arrangement.

Of all broad-leafed forests, the tropical rain forests are the outstanding examples of interdependence between plant and animal life, and the large, luscious character of many rain forest fruits are one result. A very high proportion of the commonly seen and commonly eaten fruits of the tropics are products of the rain forest, or of domestic cultivars of rain forest plants. Of these, however, only the banana and perhaps the mango have travelled worldwide, joining apples and oranges in the 'fruit you see everywhere' category. Characteristically, the stay-at-homes are those which last least well off the tree and whose seeds have little or no dormancy period. In the ever-wet forests, the ability to retard germination has no survival value; speedy processing from ripe fruit, announced by a heady aroma or bright colour, to new plant is everything, and the bigger the fruit, the bigger the animal assistant required to complete the cycle.

It is probably for this reason that the strange and distinctive habit of flowering and fruiting on the lower branches, the trunk or even just below the ground has evolved among some rain forest genera—the jackfruit, the durian and some species of fig being common examples. In these positions, the fruits will be easily accessible to the larger mammals. The distance the water supplies would otherwise have to travel to produce something the size of a jackfruit might also have been a factor: botanists can only speculate on the 'reasons' for this typically rain forest phenomenon.

Durian or Duren

In South-East Asia, the heavily armoured and strong-smelling durian is the king of rain forest fruits (Plate IV). The tall, handsome durian trees are quite easy to spot in the forest, even when not in flower or fruit, by their light-coloured bark and dark olive-green leaves, coppery or silvery on the underside; when their fruit is ripening, the aroma exuded will advertise the tree's presence for miles around. Indigenous tribesmen will camp around such a tree to prevent the animals, particularly elephants and orangutans, from feeding on it, and to be there when the fruit reaches its moment of perfection. For all its solid appearance, durian is an ephemeral fruit which must be eaten just as it is ready to split open or very soon after; connoisseurs say that even two or three hours later the taste is quite different, and the unkind remarks foreigners make about custard passed through a sewer (quoting Alfred Russel Wallace) or rotten cheese reflects how quickly the fruit spoils. There will obviously never be durian-boats as there are for bananas, and though durian-flavoured ice cream, jam, sweets and even durian chewing gum are popular in South-East Asia, these are made for durian addicts only and resemble durian to the degree that strawberry pop resembles the fresh strawberry, which is to say in nothing except the taste and smell of the overripe fruit.

Mangosteen (manggis)

If the durian is king, then the mangosteen (Plate V) must be the queen. South-East Asians believe that mangosteens should be eaten directly after durian, to cool the blood that durian heats, the perfect partner. But while durian is an acquired taste, mangosteens please everyone from the first melting mouthful of their succulent white segments, and it is a pity they are even less frequently met with outside the tropics than the durian is. Lacking the durian's formidable armour, they are also a lot easier to eat and less hard to judge for quality and ripeness. Beware the juice, however. The rind is rich in tannins—an infusion is said to be good for dysentery or fever—but what the juice most certainly will do is stain your clothes, though the brown dye will only appear after they are washed.

The cultivated species grows well in the tropics wherever rain forest, or at least moist evergreen lowland forest was the natural vegetation, and mangosteen-like fruit are produced by other *Garcinia* species fairly common in South-East Asian rain forests. But the seeds of these slow-growing trees have—as usual with rain forest trees—little dormancy, and mangosteens are also difficult to graft onto hardier stock in order to widen their natural range.

Mangoes (*mangga*)

The many varieties of cultivated mangoes make the mango (Plate VI) the best-known rain forest fruit of all apart from the banana. Mangoes have adapted to living in monsoon climates and even in city conditions while retaining the densely-crowned look of a typically shade-loving tree. Along the road they can always be recognized by their thick, relatively short trunks, their great stature in maturity, and their long thin very dark green leaves of which the new growth characteristically hangs in limp tassels of violet or pale greenish pink, depending on the species. Mangoes in flower produce purple or green upright flower panicles; when the fruit develops it hangs down on long stalks. If the rainfall is sufficient, the crop is generous and there are usually enough seasonal surpluses even in populous Asia for the preparation of all kinds of pickles, preserves and chutneys for the rest of the year.

Rambutan and Jambu

The 'hairy' *rambutan* (Plate VII), usually bright red and covered in soft spines, is a close relative of the better known litchi of Southern China, and its flesh is similar in taste and consistency. The trees are low in stature, rather untidy-looking, with big, light-coloured leaves; the fruit grow in stalkless pairs or clusters along the twigs. In season, roadside vendors surrounded by baskets of these handsome fruit are a common sight, and purchasers buy them tied up in bundles, like over-sized bunches of grapes, a temperate climate fruit the *rambutan* also resembles.

The *jambu* fruits, of which there are five commonly eaten species, are all relatives of the clove (*Eugenia aromatica*), allspice

(*Pimenta officinalis*) and the guava (*Psidium guajava*). There is *jambu bol*, *jambu mawar*, *jambu air* (*ayer*), *jambu jambolan* and *jambu air rhio*; the guava (from tropical America) is *jambu batu*. The best known is probably the red, bell-shaped *jambu air*, *Eugenia aquaea*, a crisp and thirst-quenching substitute for apples. They would probably be sold more widely if they did not bruise so easily, and if they were less attractive to ants, some of which always seem to remain hidden in the cavity at the base of the fruit in spite of all efforts to dislodge them.

5
Trees of the Monsoon Forest

'MONSOON' forest grows in those parts of the tropics where there are marked seasonal differences in rainfall; the climate of most of continental southern Asia is a monsoon climate, dry as dust for months then deluged with rain at regular times of year, though the word 'monsoon', from the Arabic *musim* or season, does not actually mean either the moisture-laden winds or the rains themselves. Plants native to this zone will flower and fruit predictably, and will often be deciduous, and generally their seeds will have a dormancy period sufficiently long to enable them to survive at least until the next rains. Such plants are altogether hardier than most rain forest species, are far more tolerant of man-made environments and make better plantation trees, the teak being one example. For all these reasons, the more decorative or useful of them have travelled widely, with man's assistance, throughout the tropics, and the visitor to the West Indies is likely to meet there many of the trees and herbs he saw in Thailand or Malaya and vice versa. In the category of the flowering trees, many may only come from as far away from South-East Asia as India, a major contributor; on the other hand, they are just as likely to be from tropical America or even Madagascar. Usually the reason for their popularity is their flowers—gardeners are the same everywhere and actively seek 'exotics'—but sometimes it is their growth style which makes them good trees for avenues, spreading blessed shade over the tropic-hot pavements, a little bit of forest coolness in the town. In any case, it would be pedantic to point the purist's finger at these well-naturalized immigrants, doing so well so far from home, and exclude them from a book like this.

The Saman or Rain Tree

One of the most common wayside trees in South-East Asia is the *saman* or rain tree, *Enterolobium saman*. Originally from South America, by the end of the nineteenth century it had spread throughout the tropics, arriving in Malaya in 1876. Planted chiefly as a shade tree because of its generous crown which can be 20 to 25 m across, the *saman* is a perfect example of the so-called 'umbrella' style of tree architecture. Because the reproductive parts need the stimulus of strong sunlight, the side branches of such trees grow faster than those in the centre, lifting the flowering parts up to the crown and outer surfaces. The result is a pleasing umbrella-shape, offering a fine display of colour in the flowering season and a generous amount of shade throughout the year.

Saman have a long flowering period, spaced over two months, during which they are covered with small, delicate pompoms, each flower (fig. 2) a mass of white stamens banded with a deep pink, surprisingly delicate blooms for such a dark, massively limbed tree. Because they flower for so long, the fruit, a long dark pod, appears long before the flowers fade. Like the West Indian carob bean, the pods contain a sweetish pulp which both children and cattle are fond of, but they are not usually collected for food, at least not in Asian cities: the curious patterns so often seen on the streets of Rangoon (for instance) are where the pods have fallen and have been pressed into the sun-softened tarmac by passing cars.

The *saman* is a fast grower, like many leguminous species, and its wood is correspondingly soft and relatively useless—a blessing in countries where so many of the other wayside trees have been cut down for various purposes and not replaced. Because it grows so large and offers so much shade, a *saman* will also provide a home for numerous epiphytes which would not otherwise survive in city conditions. Its shaggy bark provides an

Figure 2. *Saman* flower and leaves

ideal root-hold for orchids and ferns, and the deep fissures in it help to retain rain water; the tree conserves the moisture in its own leaves by closing them at nightfall, or earlier, if it is a cloudy afternoon.

The Flamboyant

The flame tree or flamboyant, *Delonix regia*, is also an umbrella-shaped tree, though not nearly so massive structurally as the *saman*. Indeed, the flamboyant spends much of the year looking indifferently bare, only to blossom out in brilliant scarlet blooms a month or two before the rains are due. The flowers (fig. 3) are big and showy, with four upright scarlet petals and a

Figure 3. Flamboyant flower

fifth, the standard, towards which the stamens bend, of mixed scarlet and white. The standard generally fades after the first day and falls off, and is therefore usually missing from blooms picked up from the ground. Discovered by a French botanist in Madagascar in 1824, it is now one of the best known, most easily recognized wayside and garden trees in Asia, and it is difficult to believe there was ever a time when it was not here.

The Tamarind

The tamarind (Plate VIII), an Indian species, is another well-travelled tree, but will not grow in places with too heavy a rainfall for them even though it is seasonal. The name is an Arab one, meaning 'Indian date'. Though the pulp surrounding the seeds is tart rather than sweet, it is sold in the markets very much the same way that the poorer quality of dates is sold in Arabia: in a sticky wad, and by weight—and like dates it is dark brown when ripe. Tamarind pulp contains a very high proportion of tartaric acid and is a favourite flavouring for syrups, jams, jellies or in fish dishes and curries, being both a preservative and tasty. Tamarind balls are regularly fed as a tonic to captive elephants, who relish them; the pulp is also useful for cleaning brass and copper utensils and ornaments. For these reasons, every village in South-East Asia will try to make room for a tamarind or two, in spite of the size they attain in old age: over 40 m high, with trunks 2 m or more in diameter.

Like *saman* leaves, tamarind leaves are bipinnate and also close at night, but the leaflets are much finer and the tamarind does not have the *saman* tree's spreading crown.

Cassias and Lagerstroemias

In general, the Cassias are fairly quick-growing and short lived leguminous species, rarely more than 10 m high, and most do not make good avenue trees because of the necessity of frequently replacing them. Nevertheless, one often does see them by roadsides and they have always been popular garden trees because of their show of flowers. *Cassia siamea* and *Cassia fistula* have yellow flowers; the latter, called the Indian laburnam, produces a mass of pendulous yellow blooms punctuated by 30 cm-long pipe-like blackish pods. This Cassia, as a matter of fact, is quite slow-growing and the timber is good. *Cassia nodosa* and *Cassia javanica* are pink-blossomed, *C. nodosa* flowering all along the branches. *C. javanica* is pictured on Plate IX. The source of the common laxative senna is usually *Cassia angustifolia*, though the bark and pods of other species also have medicinal qualities.

The Lagerstroemias, either *L. speciosa*, the 'rose of India', Plate IX or *L. flos-reginae*, the queen flower or *bungor* (fig. 4) are as frequently planted as ornamentals as the Cassias, and their rose-purple or lavender blooms compliment the Cassia's yellows and pinks. In the forest, these trees will grow to 30 m or more, a magnificent sight when covered in blooms. In the open, they usually only reach 15 or 20 m.

Figure 4.   Flowerlet of *L. flos-reginae*

Hibiscus and Mussaenda

Few South-East Asian gardens lack one or another of these handsome shrubs (figs. 5 and 6). The hibiscus, originally from China, is now pan-tropical; colourful and sturdy, easy to plant from cuttings, the single or double varieties can be grown as small or large bushes or even hedges, and seem to survive neglect, intense heat, deluges of rainfall and even droughts if they are not too prolonged.

The Mussaendas, usually small trees like the gardenia, do best in areas of high rainfall. The cultivated ones have either pink or white floral leaves; the common wild form in Indonesia is a creeper, known as the paper chase plant because of the one startlingly white leaf-like sepal beside the small orange flower.

Magnolias and Dillenias

Four-fifths of all the world's big magnolia family are native to temperate and tropical East and South-East Asia; the other fifth are American. Magnolias themselves are such distinctive trees, and so commonly planted, that they need no introduction. Some botanists claim them to be among the most primitive of all flowering plants, with their long sturdy flowers growing singly at the ends of stout twigs (in the case of Magnolias proper) and large simple leaves. The magnolia flower is similar in construction to the water lily, another 'primitive' plant which has never left the marshes thought to be the ancestral habitat of all flowering plants and were it not for their beauty, they might both be considered rather coarse-looking.

The highly perfumed yellow flowers of *Michelia chempaka* (*chempaka* is the Malay name) are particularly popular in South-East Asia as both hair ornaments and temple offerings; in Burma, the *saga wa* flowers are closely associated by legend with the most powerful of the *nat* spirits; the flowers, preserved in bottles, are always to be found by his shrines.

Figure 5. *Hibiscus rosa chinensis*

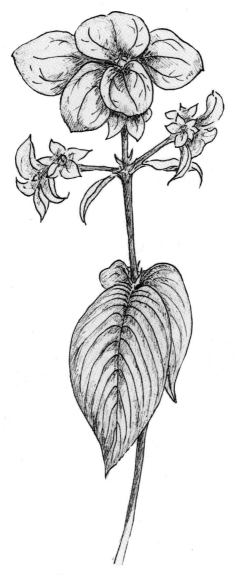

Figure 6. *Mussaenda* twig

The Dillenias are too tropical to be seen in temperate gardens, and too shade-loving to be frequently met with in towns along the hot and dusty streets. They are popular garden trees, however, and the long, dark green, glossy and markedly toothed leaves of *Dillenia indica* (on Plate VIII) cannot be mistaken for any other tree, and when it is in bloom, with its white, waxy flowers scattered among the dark foliage, it is spectacular.

Dillenia aurea, with yellow flowers and fruits (fig. 7) usually blooms just before its new leaves come out. They are trees of the lowlands, between 20 and 30 m in height, often growing along

Figure 7. *Dillenia aurea* fruit (*simpoh*)

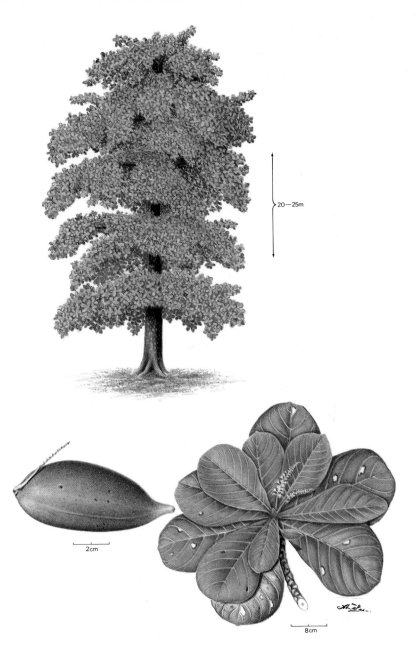

I *Terminalia catappa*, the *ketapang* or country almond

II *Barringtonia asiatica, putat laut*
Plumeria alba and *Plumeria rubra*, the frangipani

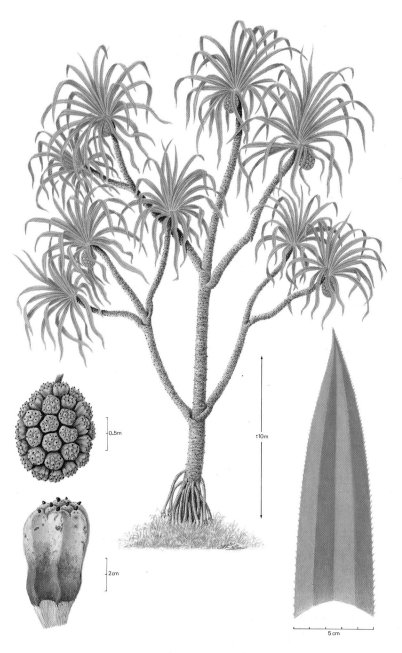

III *Pandanus tectorius*, pandan

IV *Durio zibethinus*, the durian

V *Garcinia mangostana*, the mangosteen

VI *Mangifera indica*, the common mango

4.5 cm

2 cm

VII *Nephelium lappaceum*, the rambutan
Eugenia aromatica, the clove

Dillenia indica or elephant apple

VIII   *Tamarindus indicus*, the tamarind

IX *Cassia javanica*

Lagerstroemia speciosa, bungor

X   *Cananga odorata*, the *ylang-ylang*

XI *Nelumbium nelumbo*, the Asiatic lotus

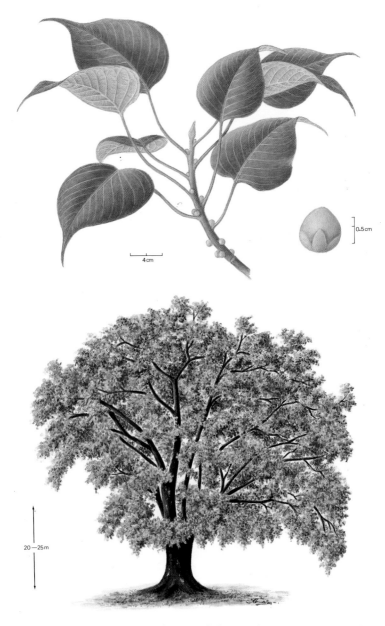

XII *Ficus religiosa*, the *bodh* or *pipal* tree

XIII *Borassus flabellifer*; the *lontar* palm

XIV *Areca catechu*, the betel nut palm

XV *Salacca edulis*, the *salak* palm

XVI *Nepenthes gymnamphora*, a pitcher plant

streams where the dropped fruits will be found in great abundance on the ground. Those of *D. indica* are called 'elephant apples'; they consist of five closely fitting sepals astringent in taste and popular in soups and curries and for making jams and jellies—as well as popular with elephants. *D. aurea* fruits are more tender, the fluted golden segments having a deliciously tart taste. *D. aurea* wood is found so frequently petrified in former swamp land in both Burma and Indonesia that the name used for petrified wood, *simpoh*, is also the name of the tree.

Ylang-ylang

In monsoon climate zones, one can also find plants which are rain forest in origin but which have adapted to seasonal variations and which are therefore more seasonal than is usual for their wild relatives, the cultivated mango being one example. There are also rain forest species which can be grown, like cloves and nutmegs, as plantation crops if the rainfall is high enough, and all the cultivated bananas of the world have been bred from two South-East Asian species (*Musa acuminata* and *M. balbisiana*). Gingers, turmeric and cardamom also adapt well to seasonal rainfall conditions, and the wild relatives of all these plants are all denizens of South-East Asian forests.

Ylang-ylang, from the Philippines, is another example of a rain forest 'escapee' now planted widely for its sweet-smelling flowers, the basis of many a local perfume industry from the Comorros to the Pacific. A distinctive-looking tree (see Plate X), it can reach a height of 30 m or more, but is commonly kept short in plantations so the workers can reach the flowers more easily. Pale green in colour, borne in clusters, they must be picked in the very early hours of the morning before the heat of the sun dissipates their aroma.

The Philippine name means 'something that flutters'; in Java the tree is called *cananga* as in its Latin name.

6

Timber Trees: The Dipterocarps and Golden Teak

WHEN people talk of South-East Asian hardwoods, it is usually the trees of the tall dipterocarp family that they mean. It is a big family with many species, but *meranti*, the name for one group, is probably the most common trade name. Strictly speaking, these trees are African in origin and were carried to Asia on the drifting parts of ancient Gondwanaland which became India. But this was 40 million years ago, and it is in South-East Asia that the family has prospered and proliferated. There are now 500 known species of which 380 are native to Malaysian and Indonesian rain forests, and it is the dipterocarps that make these rain forests the tallest in the world. Their cauliflower-like crowns, 40 m from the ground, tower over the rain forest canopy—they are easy to pick out from the air—and their straight boles of 30 m or more are a logger's delight (fig. 8). Since the last World War, the owner countries have been exploiting them intensively, and by the end of this century, at the present rate of cutting, most if not all commercially loggable stands will have been finished.

They will not grow again. The trouble is, these giants are so slow-growing that they may only flower and fruit for the first time at the ripe age of sixty, and then they do so very irregularly and infrequently thereafter. This behaviour hardly fits in with 15 or 20 year cutting cycles. They also tend to grow primarily in lowland rain forest (below 800 m) which is the very part of the rain forest most easily accessible to loggers, most suitable for roads, most desirable agriculturally or for settlements of other kinds like transmigration schemes, and therefore the part most frequently converted, accidentally or intentionally, to other

Figure 8. *Dipterocarpus alatus*, a remnant of the high forest in Rangoon

uses—whether wisely or not. Where a tree has seeded, regeneration is good; but where the mother trees no longer exist, or too much forest has been destroyed and little shade remains, there will be no more dipterocarps.

The dipterocarp family name comes from the Greek for 'two-winged seed', as in figures 9 and 10, though other species have three or even five wings or strap leaves. Characteristically, these remain attached to the seed, and as it drops from the tree it turns, heavy seed end downwards, and the 'wings' give direction to its long fall to the ground below.

While the dipterocarps as a family share many attributes which make most of the species 'good' from the timber merchant's point of view, the well-known golden wood of teak only comes from one species, *Tectona grandis*. Teak is moreover native to monsoon climate forests rather than rain forests, which make its habits more regular and the species adaptable to plantation culture. In Java, the growing and processing of teak has become a way of life in those districts where it grows well, and in Burma much of the country seems involved in one way or another with teak production. Ownership of teak trees, once a royal prerogative, is now the Burmese Government's, and most of the logs are exported to obtain hard currency; teak is second only to oil as a contributor to Burma's GNP.

Teak prefers the southwestern slopes of dry hill forests, and will grow rapidly where conditions are favourable, though attempts to extend its natural range have not been successful. When fully mature (in 70–80 years), it is a fine straight tree 26 to 30 m high with a girth of 4 to 5 m. In flower, the crown appears lacy with the upright racemes of creamy white flowers contrasting oddly with the extreme coarseness of its huge leaves. These leaves (fig. 11) are dropped seasonably and in the forests are cursed by hunters for the brittle noisy litter they make. They are also very inflammable but foresters do not object to annual fires in teak forests. Started early in the season, they

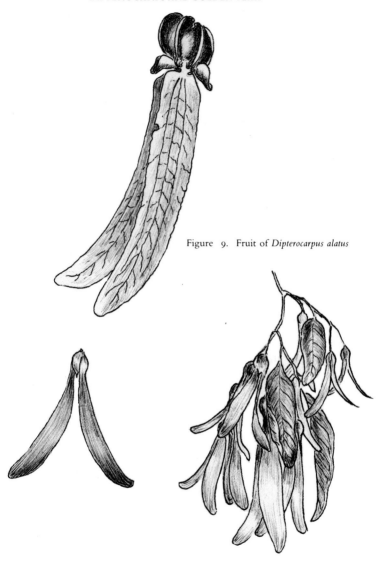

Figure 9. Fruit of *Dipterocarpus alatus*

Figure 10. Fruit and twig of *Hopea odorata*, another dipterocarp

prevent a great mass of litter from building up and further-more aid teak regeneration by cracking the extremely hard shells of the seeds. In plantations, however, the leaves rarely accumulate because they are collected and even picked from the lower branches for sale in the towns. As they are often half a metre long and almost as wide, they serve, like banana leaves, as the plates and plastic bags of South-East Asia. One will hold a full-scale Asian meal of rice and curry, while in the markets anything wet or messy or granular will get weighed out onto a teak leaf, then deftly bundled and tied with a strip of bamboo fibre.

The wood has attained worldwide popularity. The resinous oil it contains, often extracted and sold separately, renders the timber supple and easy to work as well as insect-proof and durable in damp or wet conditions, making it ideal for house construction and shipbuilding, though its handsome golden colour and usually straight grain makes it popular with

Figure 11. Mature teak leaves

furniture makers and woodcarvers as well. The intricately-carved palaces of Burmese kings were made entirely from teak, as were the older monasteries; whole trees were always used as the pillars of these big structures, and a monastery's size is still rated by their number. A village, for instance, would be proud to have an 80-pillar monastery, but a town would expect to afford one with 300 or more.

7
The Medicinal Spices

IT is more common in Eastern Asia to consider the highly aromatic plants or parts of plants as medicines rather than as flavourings, though the European taste in these matters has had some influence. The lowly turmeric root, cheap and plentiful, is a case in point. Although it is the essential ingredient of many curry powders, the base in which other flavours are compounded to taste, it is also believed to be a blood purifier, an antacid and antispasmodic, calming digestive troubles and the nerves. It is even applied externally to sprains and bruises, and rubbed on the belly after childbirth. It is also, of course, a popular yellow dye.

Wild turmeric (fig. 12), ginger, and cardamom, all members of the same family, are shade lovers and are relatively common understorey plants in both rain and closed canopy monsoon forests. Being able to send up many shoots from their creeping rhizomes, they can be found in quite large patches; the traveller, for half an hour's digging, can get himself a supply of roots for several months' use.

Wild cinnamon and nutmegs, recognizable by the aromatic oils in their leaves, by their sap or by their distinctive fruits are also common in South-East Asian forests; since the commercial species are now so widely grown in plantations, they have become more familiar to the traveller. Rows of young clove trees are particularly a common sight in Java and Sulawesi (Celebes), where their growth is much encouraged by the government—a far cry from the repressive measures by which the Dutch retained their former monopoly (Plate VII); now many a hillside is covered with these conical trees which look, when young, like pale leafed laurels touched with the attractive 'flush' of rosy red of the young growth (see Plate VII). In old

Figure 12. Wild turmeric, showing roots

age, a clove tree becomes a stately creature some 20 m high with a whitish bark—quite different in appearance.

Nutmeg trees, on the other hand, though also conical in shape, are trees of the lower storey of the forests and will be harder to identify unless you find a female in fruit. The fruit, or, rather, the nut inside is so distinctive, however, that even the tiny wild nutmegs are easily recognizable: the fruit splits open when ripe, to reveal a hard-shelled nut entwined in a red or orange net-like aril, which is the mace. Even if the mace has dried and dropped off, its impression can always be traced on the shell. Nutmeg trees also have a pinkish sap which flows freely when the thin bark is slashed and dries to the colour of blood. For this reason the Malays have given the name of *pendarah* ('blood-bearing') to the family which in turn confers on it the connotations of strength and magic. More prosaically, nutmegs are said to soothe headaches and rheumatic pains, just as oil of cloves is a palliative for toothache.

8

The Sacred Lotus and the Temple Trees

The Lotus (bunga padma)

In Hindu belief, the lotus is the life-giving principle, the symbol of the female generative organ or *yoni*, and the male generative organ or *lingam* is usually depicted with its base surrounded by lotus petals. Closely associated with gods and saints, as the lotus often is, this directly carnal symbolism may seem strong and earthy when compared to the sexless conventions of Christianity but it in no way diminishes the spiritual aspect of creation to a Hindu. Brahma is said to have been born from the lotus in (or issuing from) Vishnu's navel: Vishnu, the lotus-navelled, and Lakshmi, Vishnu's consort, even more ethereally emerged from a lotus which sprang from Vishnu's forehead, as Athena sprang from the forehead of Zeus. Nor do the carvers of countless statues of Buddhas and bodhisattvas mean to infer that those who have so far liberated themselves from carnal desires still remain seated comfortably within the female *yoni*. The familiar lotus base in this case can be regarded as a vehicle or support. Though it does not float upon the waters like a water lily, it is water-borne, growing from the depths to lift its flower up into the sunlight and air as the bodhisattvas grow toward enlightenment from their roots in the dark world of men, and as the flower itself rises clean and beautiful from the muddy waters in which it grows.

The lotus is also a widely-travelled decorative motif whose earliest appearance is in the wreath of lotus flowers adorning the head of an earth mother figure in Mohenjo-Daro, dated about 3000 BC. In 700 BC the Persians conquered Egypt, bringing with them the decorative motif if not the plant. A stone lotus flower and bud frieze decorated the palace of Ashurbanipal: Chinese vases of the Song period are often in the form of a semi-closed flower with the wavy-edged leaf as the lid.

There are only two species of lotus, one in the old world of South-East Asian origin (Plate XI) and one in the new, and it is in fact a relatively primitive plant form. Like water lilies, the water buttercups (*Ranunculus*) and the water plantain, they belong to the 'dicot' or two seed-leafed group of the flowering plants but unusually for this group do not display secondary thickening of their stems. They are the water-borne versions of the thick-trunked ferns and palms, and if the evolutionary trend has been away from the lowland swamp towards ever more refined adaptations for living entirely on land, then the freshwater plants which remain behind can be regarded as primitive failures, however beautiful their appearance or spiritual their associations.

The seed head of the lotus is unique in the plant world. The cup-shaped torus is perfectly flat on the top with the (edible) seeds pocketed in it. Below the seeds there are also enclosed air spaces which enable the seed head to float if necessary. The stalk, also edible, makes an asparagus-like vegetable, and from the rhizomes an easily digestible starch like arrowroot can be prepared. All the tissues contain alkaloids regarded as both tonic and stomachic (antacid), and one often sees bundles of stems and seed heads coiled up in the markets ready for sale, religious associations notwithstanding.

The Bodh Tree or Pipal

Siddhartha Gautama, the Buddha, entered the state of nirvana in the shade of a fig tree in Maghada, and this species of fig, the pipal, is known as the bodh tree or tree of enlightenment; in Latin, *Ficus religiosa* (Plate XII). Planted wherever there are temples or wayside resting places, it is easy to recognize because of its heart-shaped leaves with their distinctively elongated tips. It is a tree native to central and eastern India, but since it has the ability to grow either on its own trunk or as a strangling fig, it has spread widely throughout South-East Asia. Figs are favourite foods of birds or monkeys, who drop them on the branches of other trees and even on the cornices and drains of city buildings to the inconvenience of the religiously-minded: no one will wilfully destroy a fig, most particularly this species, and a non-believer must be called in to perform this task lest the tree take real hold and begin to break up the building.

When grown as a planted tree, the trunk is never more than 4 to 6 m high, breaking up into a many-branched and rather untidy crown. In old age, the lowest branches will grow very long and almost horizontally, in danger of breaking off from the trunk because of their weight. In temple precincts, and sometimes in the forest, if a free-standing tree is growing beside a path, it is customary to prop up the oldest and heaviest branches with sticks, often in several places; there is a saying that if you prop up the old fig's branches, even so will the faith of Buddhism sustain you in your old age.

Frangipani

There seems to be no reason for planting the now ubiquitous frangipani (Plate II) in temple grounds, other than for the beauty and scent of its flowers. The three species common throughout South-East Asia are imports from tropical America, the Spaniards having first brought it to the Philippines. By the seventeenth century, frangipanis had reached Amboina (Ambon) in the Moluccas, and from there were carried to Java and Madura where its use was principally medical, as a diuretic, especially in the treatment of venereal disease, and as a remedy for skin complaints and intermittent fevers. The *Plumeria* genus is a member of the Apocyanaceae family, as *Strophantus*, *Rauwolfia* and *Vinca*. In Malaysia, frangipanis are or were more frequently planted by the sides of graves than near temples; the Malay name of *pokok kubur* reflects this. Its present-day appearance in so many gardens is probably the result of unsuperstitious European taste in such matters and the fact that it is so easy to grow from any branch broken off and stuck in the ground.

9
The Grasses: Bamboo and Rice

GRASSES are distinguished from reeds by the regularly spaced closed sections in their normally hollow stems, and it is from these nodes that new growth generally sprouts. This structure is particularly evident in bamboos, the giants of the grass world, and is one of the reasons they are so outstandingly useful. South-East Asian forests seem to provide a size of bamboo for every purpose: water or toddy carriers, rice steamers, storage 'jars' or easily-fashioned mugs or water boilers: no need to carry a kettle in forests where there are bamboos. The hollow stems also combine lightness with strength when used whole as structural members in house-building, and the fact that their fibres all run lengthways makes them easy to split and flatten for use as flooring or to make the cheap form of prefabricated walling so popular in Indonesia (fig. 13). Split in half and cut across, they make handsome tiles for roofs, though more popular in the drier islands because of the unsightly mould which blackens them in areas of heavy rainfall. Better than the tiles is the Balinese form of bamboo thatching, which consists of fine bamboo strips bent over battens (as coconut leaflets are in East Africa). When the battens are placed in close rows the roof is a thing of great beauty seen from inside and will withstand the tropical downpours for fifty years.

Bamboos are generally easy to grow, adaptable and able to survive most adverse conditions and forms of interference, one of the most devastating being the presence of elephants. Elephants are inordinately fond of bamboo and think nothing of reducing a whole grove to something that looks like the aftermath of a hurricane in one night's feeding time. But in fact, bamboos seem to thrive on disturbance and the opportunities for new growth it creates, and extensive bamboo brakes are a common form of secondary jungle.

This excellent grass does have one quirk, however, and the current plight of the giant pandas in southwestern China has made nearly the whole world aware of it: this is the seemingly curious habit of sudden gregarious flowering when all the bamboos of one species for miles around, acting on some internally or externally triggered impulse, will mobilize their resources into producing flowers and seeds after which they die. If this were an annual or even a regular event, the giant panda as a species would long since have learned to supplement their diet with other plants. It is, however, this same faculty—as E. J. H.

Figure 13. A bamboo matting house wall being carried

Corner points out in *The Life of Plants*—that makes grass species such efficient seed producers, and the trait for which they are bred commercially.

The world depends on only a few of thousands of grass species to provide its staple starches, and rice is one of the oldest cultivated grains. On the evidence of the number of wild species it may have been introduced to China from India, where there are five or six thousand types of named rice, but no one knows whether the early cultivated varieties were derived from the so-called hill rice of dry zones or from the plant of lowland marshes. There is evidence for both, and genetic differentiation is still going on in the wild—as it is being fostered in the giant laboratories devoted to the purpose such as the Rice Research Institute in the Philippines. Attempts are made genetically to control the maturation period, the length and strength of the straw, the ease with which the husk is shattered, the plant's hardiness, yield and resistance to salinity and insect attack, the seed's dormancy and its nutritional quality. Unfortunately, through genetic linking, some of the least hardy varieties have the highest yields, and those with the shortest ripening time (desirable) tend to have seeds which sprout immediately on maturity (undesirable), but many improved varieties have been introduced throughout Asia which have doubled or tripled yields where conditions have been favourable.

With so many millions of people depending on rice as a staple, it is natural that there would be many superstitions and folklore surrounding its cultivation. The belief that rice has a soul, that this soul is contained in the best-formed seed and that this seed should be kept for the next year's planting may only be a spiritual expression of good agricultural practice, but there are still many otherwise sophisticated modern Asians who believe that rice tastes better if it is dehusked by hand in mortars rather than by machine—a faint whisper of the belief that rice, the mother, should be treated respectfully.

10

The Palms

Is South-East Asia a bamboo culture or a palm culture? Since nearly every locally-produced artefact, utensil, house component, rope, basket or piece of furniture seems to be made out of one material or the other, it's hard to say. But palms also provide fruits, vegetable matter, sugar and yeasty drink, and in the southern, wetter areas they must win because of the immense variety of species. Tropical Asia is the 'palm centre' of the world, with half again as many species as are found in all of tropical America, and over ten times as many as in Africa, and they range in life-style from the highly adaptable, like the ubiquitous coconut and the stately palmyra, to the palm specialists of the swamps (sago) and deltas (nipah) and the snaky-stemmed palms of the tall rain forests, the versatile rattans, for which South-East Asia is the world's main source.

Borassus, the Palmyra Palm

This tall, conspicuous palm with its neat circular crown of fan-shaped leaves (Plate XIII) is characteristic of a seasonally dry climate. Aside from the coconut, no palm is more common and none is more widely cultivated. It is also one of the hardiest, as tolerant as the coconut of poor growing conditions and replacing it at altitudes at which the coconut will not grow. Massive and often solitary, its big trunk zigzagged with the scars of dropped or severed leaves and, in season, bearing its distinctive three-chambered edible fruits, the palmyra dominates the landscape wherever it grows, whether silhouetted against the night-time illuminations of Rangoon's Shwe Dagon pagoda or rising out of the pale grass of Komodo Island to form Komodo's distinctive skyline.

As is possible with several species of palm, a toddy (fig. 14)

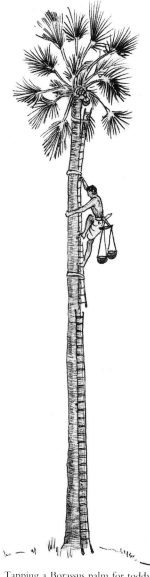

Figure 14. Tapping a Borassus palm for toddy in Burma

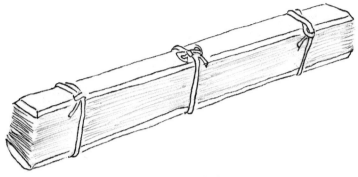

Figure 15. A palm leaf book

can be obtained from the unripe inflorescence which can be drunk fresh, evaporated to make sugar (jaggary), distilled to make spiritous liquor or allowed to sour into vinegar. The mature leaves thatch houses, the leaf stalks provide fibre for rope and the ribs of the leaves are made into brooms; in all this, the palmyra is no different from the coconut. But the leaves of the *lontar*, to give the palm its Malay name, are most valued for their use, when immature, as the narrow 'pages' of sacred books in which Buddhist and Hindu scriptures have been recorded, the shape and size of these books being strictly governed by the maximum length and width of usable leaf blade (about 6 by 60 cm). Inscribed with a stylus, rubbed with charcoal dust to make the letters visible, the leaves are then pierced and tied together and the whole protected by handsome wooden covers, often with lacquer designs (fig. 15).

Areca, the Betel Nut Palm

In nearly every southern Asian village, given sufficient rain, there will be at least a line or two of very thin, very tall and very elegant palms with tufts of graceful leaves at the top. This is the *Areca* palm (Plate XIV), which the Indians call 'an arrow shot from heaven'. In season, a spathe or two of bright orange-red fruits will appear; each fruit, about 5 cm long, encloses the nut which is the essential ingredient of *betel* or *pan*, the popular chewing mixture of tropical Asia.

The habit is surprisingly soon acquired in spite of the complication of several necessary ingredients, the copious red-stained saliva they induce and the fact that the steady user will end up with teeth reduced to blackened stumps in a permanently carmined mouth. Strangers are often shocked by the effect of premature ageing this has on some once-attractive woman, as well as by the swatches of red spittle on walls and pavements, and many modern Asians regard these side effects with equal abhorrence. It is not the palm nut which rots the teeth but the slaked lime which, with a leaf of the *Piper betel*, a pepper family vine, are the other two essential ingredients. To these are added some cutch or gambier, gum of the *Acacia catechu* tree, spices to taste, and often tobacco. Each village betel stall will have its ingredients spread out for the buyer to choose from, and a host's expectation that every guest will want to chew betel has given rise to as much paraphernalia as tobacco-smoking and snuff-taking has in western countries, or the inhaling of tobacco smoke through water has among the Arabs. In Burma, the betel-box is an elegant cylindrical lacquer container with three internal trays, the whole intricately decorated with traditional designs, those from the old palaces being particularly large and handsome. Other countries go in for small personal boxes of brass or silver; the ornate Achinese boxes with their cutting and spreading implements attached to them by delicate chains are also works of art.

Arenga, the Sugar Palm

The dark, gloomy-looking sugar palm is the complete opposite of the slender *Areca* in appearance. Thick-trunked, usually squat, its dull dark green leaves coarsely feathered and an untidy collection of black leaf sheath 'skirts' swathed around its trunk, it is not a palm to be planted for show in an avenue or as an ornament to the latest building. But appearances are not everything. These same ugly leaf sheaths are the source of an excellent vegetable fibre, very like horsehair, which can be made into rope or used as a first-class thatching material, durable and waterproof—even against sea water, very important to the builders of shrines and temples on rocky promontories: in Bali, *Arenga* 'hair' is the traditional temple thatch, much admired for its distinctive appearance as well as its other qualities.

In the production of toddy, the *Arenga* outclasses all its other palm rivals, and palm sugar from this source is cheap and plentiful throughout the wetter parts of South-East Asia. Lavish production makes these palms heavily dependent on water supplies and correspondingly sensitive to droughts.

With the fruit of this palm the Moluccans, in their seventeenth century wars against the Dutch, are said to have captured the East Javan town of Surabaya. They crushed the fruits, whose walls are full of irritant crystals, and threw them into the river upstream of the town, making the town's only water supply into what the Dutch called 'hell water' and thus forcing their surrender.

When the stout *Arenga* stems are mature, they are sometimes felled for the starch or sago they contain, as is the better known sago palm.

Whatever the palm species, the method of tapping toddy (from *tadé*, a Sanskrit word) is more or less the same. Though holes can be made in the trunks of palm from which the liquid will flow, this harms the plant, weakening it and making the entrance of boring beetles and other pests easy. The preferred method is to tap the still-closed spadix, either male or female depending on the species of palm. The tapper selects a well-developed spadix, taps it all over to bruise the unopened flowers within and binds it tightly. He then shaves off the tip daily until the toddy starts to flow, and thereafter morning and evening when the toddy is collected, only removing as much as will ensure a good flow, and will not 'use up' the inflorescence too quickly. When the toddy is used to make jaggary, it is boiled in long shallow troughs until it has evaporated sufficiently for the sugar to crystalize. Jaggary is variously light to dark brown and is usually sold in the shape of the container in which it has hardened: half coconut shells, small lengths of bamboo or plaited baskets, though in Burma the best quality is sold in lumps rolled between the palms of the makers' hands. Jaggary has a most distinctive taste, having more in common with maple than with cane sugars.

Metroxylon, the Sago Palm

The saying is that where sago is the staple starch, there you will find an indolent population. There is no need to plant or cultivate the sago palm, *Metroxylon sagu*; just remember to visit the swamps in which they grow sufficiently often to find one about to flower, when the trunk will contain its maximum of starchy fibre which the flowering will use up. After flowering, the main trunk withers and dies, to be replaced by the suckers growing liberally at its base.

Sago palms are easy to recognize by this clumping habit. A mature stem will be no more than 10 m high and about 1 m or 1.5 m in girth; they have feathery but spiny leaves of a very dark green and scale-covered seeds, like rattans and *Zalacca edulis*, the *salak* palm. In the wild they occur only east of Borneo, the most important palm species in eastern Indonesia and New Guinea.

Fortunately the palm itself provides all the materials necessary to process its pith into edible starch. The outer trunk is very hard and only needs splitting to make two excellent troughs.

One is propped at an angle and above the other in such a way that the water poured over the pith, collected from both halves, will run down one trough and into the other, where the washed out starch will collect and settle. Excess water in the bottom trough only needs to be poured off from time to time, and the starch scooped onto leaves to drain and dry. The average yield of a palm will vary between 150 and 300 kg, but exceptionally a good trunk will yield over 500 kg. It is easy to see that between three and five days' work will produce enough sago starch to feed a family of seven for three months, and a month's work will provide for the year, leaving the rest of the man's time free to fish, collect cloves, drink or run a bicycle taxi, the latter being one of the most popular occupations, at least on the island of Ambon in the Moluccas. So popular, in fact, that the local government has had to divide the number of these by half, ordering one half to be painted white and the other half red and decreeing alternate red and white days for fare-taking.

The usual form in which sago is eaten is as a hot gruel known as *papeda*. It looks a simple dish, but it requires considerable skill to pick up some of this gelatinous and gluey substance with a fork or chopsticks, then dip it in the hot fish sauces and get it as far as your mouth, where you must make no attempt to chew it but must swallow it whole, instantly. Every Moluccan has some entertaining tale of a foreigner with his jaws tightly gummed together, or gagging helplessly with his eyes starting from his head.

Salacca (or Zalacca), the Salak Palm

The *salak* is even further removed in looks from the coconut, palmyra or betel-nut type of palm. It has no stem or trunk at all and its fronds, like those of many rattans, are a mass of formidable spines or thorns (Plate XV). In the heart of this armour the pear-shaped fruit develop, delicious in spite of their prickly scaly skins and slightly fibrous texture but only when fully ripe: there is no delicate sweet jelly stage as in the coconut.

The fruit's distinctive scaly skin identifies the *salak* as a member of the same palm 'line' or sub-family as the sago, the raphia, rattans, the endemic Sulawesi (Celebes) palm, *Pigafettia*, and two New World palm genera.

Nipa (or Nypa), the Nipah Palm

The nipah, a salt water palm, is the most ancient of all palms and in fact one of the oldest flowering plants on earth. Once distributed along the shores of the Tethys Sea—fossil nipahs have been found in Europe and southern England—they are now only found in the eastern tropics, from Ceylon along the Bay of Bengal, throughout South-East Asia and as far east as Queensland and the Solomon and Mariana Islands. They are trunkless palms, and at first glance may look as though they do not have much to offer beyond their huge 7 m long feathery leaves, which make an excellent thatch. These leaves rise erect from a rosette, a primitive form of plant structure—of which the cycads, also very ancient, are another example—whose buoyant nature accounts for the palms' ability to colonize or re-colonize muddy sea shores after storms, and their creeping habit enables them to consolidate newly-formed land behind the seaward advance of the mangroves. Like mangroves, they also help to dissipate the worst effects of cyclones and tidal waves, and many shore birds find thick growths of nipah an ideal protection. The young inflorescence can also be tapped for toddy and the yield is copious. In Sandakan in Sabah (formerly North Borneo) alcohol distilled from nipah toddy was used to run agricultural vehicles before the Second World War, and although the sap does not readily crystalize into sugar, it is a cheap source since the palms require neither planting nor cultivating. In some countries (Malaya, for instance), the young leaves are cut for

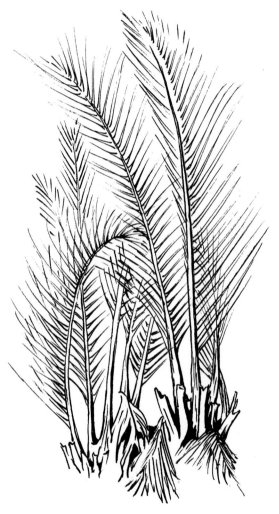

Figure 16. Nıpa palms, much cut for thatch

making cigarette papers (*rokok nipah*); what with this use and the demand for thatch, nipah palms in accessible areas usually look like the sketch in fig. 16.

II

The Basket Ferns

FERNS are sometimes regarded as the sign of degraded soils or barren land for the reason that they will manage to grow where little else can, provided only that there is sufficient moisture, even though this moisture may only be a heavy night-time dew. They are one of the oldest plant forms; relatives which, like ferns, reproduced themselves by spores rather than flowers and seeds made up the forests, 40 m or more high, of which the Coal Measures were formed. That so much vegetable matter remained to be made into coal is in marked contrast to what remains unprocessed in a rain forest, where all litter is immediately transformed into reabsorbable minerals by an army of bacteria, fungi and insects.

There are now about 10,000 species of fern in the world, ferns for every type of habitat: sun or shade, rooted in the ground or epiphytic on trees or rock surfaces, growing at high altitudes or even in water. Their fronds may be either delicate and feathery, or tough and leathery to conserve moisture; their spores are contained in the round or elongated brown structures on the reverse side of the leaf.

The bird's nest fern, *Asplenium nidus* or *paku langsuyar* belongs to one of the world's largest fern genera (650 species) and one of the commonest in Malaya. Aspleniums get their name from the supposed resemblance of the leaf to the human spleen; *nidus* refers to the nest-like manner in which the dying leaves become enmeshed with the roots to form a thick hairy tangle at the base of the plant. This spongy mass is extremely efficient at absorbing and holding water as well as the discarded plant matter which the long upright leaves catch and funnel inwards as humus.

Water retention is the biggest problem for any plant whose roots are not in the ground, and the bigger its form, the more water it will use. Most of the large epiphytic ferns, therefore, have thick, firm leaves with a leathery cuticle and Asplenium leaves are no exception.

The other common type of basket fern, the stag's horn fern *Platycerium coronarium* (*semun bidaduri* or *rumah langsuyar*), has equally thick but more specialized leaves. The erect ones are about 60 cm long and very nearly as broad, more efficient collectors of compostable material than the narrower *Asplenium nidus* leaves. The others, which hang downwards, are the ones from which the plant gets its name. As these mature and die off, they curl inwards, holding and pressing the accumulated debris against the roots and other leaves to form a tightly laced container wherein new roots can find nourishment and new leaves form. Only these 'nest' leaves are fertile. Though normally only about 2 cm wide, a nest leaf will divide unequally at first to form a kidney-shaped lobe which is a velvety mass of spore receptacles and small hairs; it will then continue growing in the distinctive stag's horn pattern until it reaches a length of 2 m or more. Young plants will at first consist only of nest leaves in a circular or crown-like form.

Both these large ferns are such efficient water and mineral collectors that you will often see several other species of epiphytic ferns, orchids or mosses growing just below them, benefiting from their surplus.

The Elephant Creeper and the Pitcher Plant

ALTHOUGH lianas are more plentiful in true rain forest, where there is more of everything, it does not follow that monsoon forests are without their vines, or that some of these specialized growth forms whose structure is devoted singlemindedly to the task of raising water from ground level to their flowering and fruiting parts in the forest canopy are necessarily smaller in scale away from the everwet rain forest. The 'elephant creeper', *Entada phaseoloides* (fig. 17) is one of the biggest vines in the world and certainly produces the largest seed pod, 2 to 3 m long and containing eight or nine large, very hard and smooth-shelled beans. The pods are usually dangling so high up in the tree support that many people do not at first make the connection between them and the twisting, convoluted stems on the ground.

Almost every part of this remarkable plant has its use. The fibrous bark of the stem is used for making cord or, flattened and beaten out, as a saddle blanket for baggage elephants. Pieces of the creeper are cut off, mashed and used as a soapy pad to wash the elephants after a day's work: all parts of the plant contain non-toxic saponins, and elephants, whose skins are extraordinarily sensitive to friction, insect attack or the pressure of an improperly balanced load or frayed harness, must be scrubbed daily. The elephant creeper, usually found growing along river banks, is ready to hand for the evening bath. The seeds are often pounded to obtain a good shampoo for human beings, or to be used internally for their worm-killing, tonic or slightly emetic properties. The juice extracted from the wood and bark is applied to jungle sores or tropical ulcers. The dried

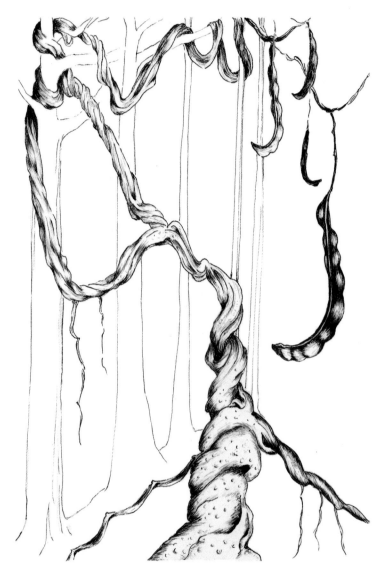

Figure 17. Elephant creeper, *Entada phaseoloides*, in the forest

and hardened seeds are used for polishing handmade paper and burnishing pottery, and are also prized for playing the highly competitive game of *gonyin* (the name of the plant), the rather violent Burmese version of conkers. Pounded, the vine is also used like derris root as one of the less harmful fish poisons, of limited duration—unlike such agricultural poisons as Endrin, widely used for this purpose by those unaware of its effect on the water and the wildlife.

The pitcher plants, *Nepenthes*, are another interesting form (Plate XVI). Tropical forests are filled with peculiar adaptations to generally poor soil conditions. Where for some reason the soil is badly leached of its mineral content, like the sandy *kerangas* heaths in parts of Borneo, or in the rocky craters of extinct volcanoes or on the mossy walls of cliffs, you are likely to find some of these extraordinary plants which supplement their meagre mineral supplies by trapping and digesting a steady catch of unwary insects. One or several leaves on the plant will develop specialized ends, called pitchers from their shape, in which a nectar is formed, attracting ants and other insects to the lip of the pitcher and then inside, where a slippery secretion prevents them from escaping again.

The Malays call this plant, at least the larger-pitchered species, the 'monkey's cooking pot', and the Dayaks of Borneo, among other people, are fond of steaming their rice in these elegant containers, which impart both a delicate colour and flavour to it—besides making the meal easy to carry. Fibres for rope- or basket-making can be obtained from the stems of some species, a more mundane use. In Borneo, where pitcher plants are most plentiful and where, in that land of high rainfall, they can be found from sea level to 800 m, one tends to take their presence for granted and treat them almost as weeds, forgetting how very endangered some *Nepenthes* species are.

Glossary

Bipinnate. A kind of compound leaf consisting of opposite or alternate subdivided leaflets on either side of a central axis.

Deciduous. Of a plant, the seasonal shedding of all its leaves at one time.

Dormancy. Of a seed: the attribute of retaining fertility over a period of time until appropriate conditions for germination and growth are available.

Epiphyte. A plant using another for support and therefore not rooted in the ground, but which is not a parasite.

Genus (pl. genera). Latin term for group of plants (or other living things) closely resembling each other between the larger group name of 'family' and the even more definitive group name of 'species'.

Habit. The way a plant grows, its 'architecture'.

Inflorescence. The whole flowering shoot of a plant as opposed to an individual flower.

Pioneer. Of a plant, one able to colonize bare or poor soils or grow in very adverse conditions.

Rosette. In plants, a growth form in which a cluster of leaves radiates from the end of a twig or, in a 'primitive' plant form, at the top of a thick trunk.

Saponin. Soapy substance (glucoside) contained in some plants or parts of plants.

Sepal. Outer leaf-like part of a flower several of which will have enclosed the bud.

Spadix. A type of inflorescence (q.v.), a thick fleshy spike closely set with flower buds and entirely covered and protected (by a spathe).

Spore. The non-sexual reproductive unit of a fern, fungus, moss or liverwort, none of which reproduce by seed.

Standard. One petal of a flower sometimes differently coloured or more upright than the others.

Torus. A cone-shaped swollen part at the top of a flower stalk.

Select Bibliography

Burkill, I. H., *Dictionary of the Economic Products of the Malay Peninsula* (London, Crown Agents, 1935; reprinted Singapore, 1966).

Corner, E. J. H., *Wayside Trees of Malaya*, vol. I (Singapore, Government Printing Office, 1940; reprinted 1965).

_____ *The Life of Plants* (London, Weidenfeld & Nicholson, 1964).

_____ *The Natural History of Palms* (London, Weidenfeld & Nicholson, 1966).

Veevers-Carter, W., *Riches of the Rain Forest* (Singapore, Oxford University Press, 1984).

Wallace, A. R., *The Malay Archipelago* (London and New York, Macmillan & Co., 1890).

Whitmore, T. C., *Palms of Malaya* (Kuala Lumpur, Oxford University Press, 1973.)

Zaheer, S. H. (Chairman of Editorial Committee), *The Wealth of India* (New Delhi, Council of Scientific and Industrial Research, 1966).